U.S.NRC
United States Nuclear Regulatory Commission
Protecting People and the Environment

NUREG-1556
Volume 19, Rev. 1

Consolidated Guidance about Materials Licenses

Guidance for Agreement State Licensees about NRC Form 241 "Report of Proposed Activities in Non-Agreement States, Areas of Exclusive Federal Jurisdiction, or Offshore Waters" and Guidance for NRC Licensees Proposing to Work in Agreement State Jurisdiction (Reciprocity)

Draft Report for Comment

Office of Federal and State Materials and Environmental Management Programs

AVAILABILITY OF REFERENCE MATERIALS
IN NRC PUBLICATIONS

NRC Reference Material

As of November 1999, you may electronically access NUREG-series publications and other NRC records at NRC's Public Electronic Reading Room at http://www.nrc.gov/reading-rm.html. Publicly released records include, to name a few, NUREG-series publications; *Federal Register* notices; applicant, licensee, and vendor documents and correspondence; NRC correspondence and internal memoranda; bulletins and information notices; inspection and investigative reports; licensee event reports; and Commission papers and their attachments.

NRC publications in the NUREG series, NRC regulations, and Title 10, "Energy," in the *Code of Federal Regulations* may also be purchased from one of these two sources.
1. The Superintendent of Documents
   U.S. Government Printing Office
   Mail Stop SSOP
   Washington, DC 20402–0001
   Internet: bookstore.gpo.gov
   Telephone: 202-512-1800
   Fax: 202-512-2250
2. The National Technical Information Service
   Springfield, VA 22161–0002
   www.ntis.gov
   1–800–553–6847 or, locally, 703–605–6000

A single copy of each NRC draft report for comment is available free, to the extent of supply, upon written request as follows:
Address: U.S. Nuclear Regulatory Commission
          Office of Administration
          Publications Branch
          Washington, DC 20555-0001
E-mail: DISTRIBUTION.RESOURCE@NRC.GOV
Facsimile: 301–415–2289

Some publications in the NUREG series that are posted at NRC's Web site address http://www.nrc.gov/reading-rm/doc-collections/nuregs are updated periodically and may differ from the last printed version. Although references to material found on a Web site bear the date the material was accessed, the material available on the date cited may subsequently be removed from the site.

Non-NRC Reference Material

Documents available from public and special technical libraries include all open literature items, such as books, journal articles, transactions, *Federal Register* notices, Federal and State legislation, and congressional reports. Such documents as theses, dissertations, foreign reports and translations, and non-NRC conference proceedings may be purchased from their sponsoring organization.

Copies of industry codes and standards used in a substantive manner in the NRC regulatory process are maintained at—
    The NRC Technical Library
    Two White Flint North
    11545 Rockville Pike
    Rockville, MD 20852–2738

These standards are available in the library for reference use by the public. Codes and standards are usually copyrighted and may be purchased from the originating organization or, if they are American National Standards, from—
    American National Standards Institute
    11 West 42nd Street
    New York, NY 10036–8002
    www.ansi.org
    212–642–4900

NUREG-1556
Volume 19, Rev. 1

United States Nuclear Regulatory Commission

Protecting People and the Environment

Consolidated Guidance about Materials Licenses

Guidance for Agreement State Licensees about NRC Form 241 "Report of Proposed Activities in Non-Agreement States, Areas of Exclusive Federal Jurisdiction, or Offshore Waters" and Guidance for NRC Licensees Proposing to Work in Agreement State Jurisdiction (Reciprocity)

Draft Report for Comment

Manuscript Completed: June 2013
Date Published: August 2013

Prepared by:
L. Cuadrado
C. Z. Gordon
M. R. Simmons
G. M. Warren

Office of Federal and State Materials and
   Environmental Management Programs

COMMENTS ON DRAFT REPORT

Any interested party may submit comments on this report for consideration by the NRC staff. Comments may be accompanied by additional relevant information or supporting data. Please specify the report number NUREG-1556, Volume 19, Revision 1, in your comments, and send them by the end of the comment period specified in the *Federal Register* notice announcing the availability of this report.

Addresses: You may submit comments by any one of the following methods. Please include Docket ID **NRC-2013-0186** in the subject line of your comments. Comments submitted in writing or in electronic form will be posted on the NRC website and on the Federal rulemaking website http://www.regulations.gov.

Federal Rulemaking Website: Go to http://www.regulations.gov and search for documents filed under Docket ID **NRC-2013-0186**. Address questions about NRC dockets to Carol Gallagher at 301-287-3422 or by e-mail at Carol.Gallagher@nrc.gov.

Mail comments to: Cindy Bladey, Chief, Rules, Announcements, and Directives Branch (RADB), Division of Administrative Services, Office of Administration, Mail Stop: 3WFN-06-A44MP, U.S. Nuclear Regulatory Commission, Washington, DC 20555-0001. Faxes may be sent to RADB at 301-287-9368.

For any questions about the material in this report, please contact: Tomas Herrera, Sr. Project Manager at 301-415-7138 or by e-mail at Tomas.Herrera@nrc.gov.

Please be aware that any comments that you submit to the NRC will be considered a public record and entered into the Agencywide Documents Access and Management System (ADAMS). Do not provide information you would not want to be publicly available.

ABSTRACT

This technical report contains information intended to provide program-specific guidance and assist applicants and licensees in preparing applications for a general license under Title 10 of the *Code of Federal Regulations* (10 CFR) 150.20, "Recognition of Agreement State Licenses," by describing the types of information needed from the licensee to complete NRC Form 241, "Report of Proposed Activities in Non-Agreement States, Areas of Exclusive Federal Jurisdiction, or Offshore Waters." This report should be used in preparing requests for NRC Form 241; however, the guidance contained in this report does not represent new or proposed regulatory requirements. It is intended for use by Agreement State licensees, NRC licensees, and NRC staff, and it will also be available to Agreement States. This document also provides contact organization guidance to NRC licensees who wish to work in Agreement States.

Paperwork Reduction Act Statement

This NUREG contains information collection requirements that are subject to the Paperwork Reduction Act of 1995 (44 U.S.C. 3501 et seq.). These information collections were approved by the Office of Management and Budget (OMB), approval numbers 3150-0007, 3150-0009, 3150-0010, 3150-0013, 3150-0020, and 3150-0130.

Public Protection Notification

The NRC may not conduct or sponsor, and a person is not required to respond to, a request for information or an information collection requirement unless the requesting document displays a currently valid OMB control number.

FOREWORD

The U.S. Nuclear Regulatory Commission's (NRC's) NUREG-1556 technical report series provides a comprehensive source of reference information about various aspects of materials licensing and materials program implementation. These reports, where applicable, describe a risk-informed, performance-based approach to licensing consistent with the current regulations. The reports are intended for use by applicants, licensees, license reviewers, and other NRC personnel. The NUREG-1556 series currently includes the following volumes:

Volume No.	Volume Title
1	Program-Specific Guidance about Portable Gauge Licenses
2	Program-Specific Guidance about Industrial Radiography Licenses
3	Applications for Sealed Source and Device Evaluation and Registration
4	Program-Specific Guidance about Fixed Gauge Licenses
5	Program-Specific Guidance about Self-Shielded Irradiator Licenses
6	Program-Specific Guidance about 10 CFR Part 36 Irradiator Licenses
7	Program-Specific Guidance about Academic, Research and Development, and Other Licenses of Limited Scope
8	Program-Specific Guidance about Exempt Distribution Licenses
9	Program-Specific Guidance about Medical Use Licenses
10	Program-Specific Guidance about Master Materials Licenses
11	Program-Specific Guidance about Licenses of Broad Scope
12	Program-Specific Guidance about Possession Licenses for Manufacturing and Distribution
13	Program-Specific Guidance about Commercial Radiopharmacy Licenses
14	Program-Specific Guidance about Well Logging, Tracer, and Field Flood Study Licenses
15	Guidance about Changes of Control and about Bankruptcy Involving Byproduct, Source, or Special Nuclear Materials Licenses
16	Program-Specific Guidance about Licenses Authorizing Distribution to General Licensees

Volume No.	Volume Title
17	Program-Specific Guidance about Special Nuclear Material of Less Than Critical Mass Licenses
18	Program-Specific Guidance about Service Provider Licenses
19	Guidance for Agreement State Licensees about NRC Form 241 "Report of Proposed Activities in Non-Agreement States, Areas of Exclusive Federal Jurisdiction, or Offshore Waters" and Guidance for NRC Licensees Proposing to Work in Agreement State Jurisdiction (Reciprocity)
20	Program-Specific Guidance about Administrative Licensing Procedures
21	Program-Specific Guidance about Possession Licenses for Production of Radioactive Materials Using an Accelerator
22	Reserved

The current document, NUREG-1556, Volume 19, Revision 1, "Guidance for Agreement State Licensees about NRC Form 241 "Report of Proposed Activities in Non-Agreement States, Areas of Exclusive Federal Jurisdiction, or Offshore Waters" and Guidance for NRC Licensees Proposing to Work in Agreement State Jurisdiction (Reciprocity)," is intended for use by applicants, licensees, and NRC staff. This revision provides a general update to the previous information contained in NUREG-1556, Volume 19, dated December 2000.

A team composed of staff from NRC Headquarters and NRC regional offices prepared this document, drawing on their collective experience in radiation safety in general and as specifically applied to reciprocal recognition of Agreement State licenses.

This document describes methods acceptable to the NRC staff for implementing specific parts of the Commission's regulations; the process the NRC staff uses to evaluate reports; and data the NRC staff needs to review reports of reciprocity activities. The Agreement States have established comparable programs to handle reciprocity activities in their states. NUREG-1556, Volume 19, Revision 1, is not a substitute for NRC regulations. The approaches and methods described in this report are provided for information only. Methods and solutions different from those described in this report may be acceptable if they include a basis for the NRC staff to make the determinations needed to approve a reciprocity request.

Brian J. McDermott, Director
Division of Materials Safety and State Agreements
Office of Federal and State Materials and Environmental Management Programs

CONTENTS

APPENDICES

FIGURES

TABLES

ACKNOWLEDGMENTS

The working group thanks the individuals listed below for assisting in the review and update of the report. All participants provided valuable insights, observations, and recommendations.

The working group would like to thank the staff in the regional offices of the NRC and all of the States who provided comments and technical information which assisted in the development of this report.

The working group also thanks Lisa Dimmick, John O'Donnell, Monica Orendi, Tara Weidner, and Duane White for developing the formatting and language used in many parts of the report.

The Participants for this Revision:

Cuadrado, Leira
Gordon, Craig
Lenehan, Daniel
Olmstead, Joan
Simmons, Michelle
Warren, Geoffrey

ABBREVIATIONS

AEA	Atomic Energy Act 1954, as amended
COC	Certificate of Compliance
CFR	*Code of Federal Regulations*
DOT	United States Department of Transportation
FAA	Federal Aviation Administration
FSME	Office of Federal and State Materials and Environmental Management Programs
GPS	Global Positioning System
mCi	millicuries
MOU	Memorandum of Understanding
NSF	National Science Foundation
NRC	United States Nuclear Regulatory Commission
OMB	Office of Management and Budget

DEFINITIONS

For the purpose of this document, the following definitions apply:

Agreement State. Any State with which U.S. Nuclear Regulatory Commission (NRC) or the Atomic Energy Commission has entered into an effective agreement under Subsection 274b of the Atomic Energy Act of 1954, as amended.

Change. Providing information, on NRC Form 241, "Report of Proposed Activities in Non-Agreement States, Areas of Exclusive Federal Jurisdiction, or Offshore Waters," or the equivalent, that modifies items approved in the initial filing. This may include changes to the radioactive material; change of licensee address, contact, telephone number, or fax number; change of client telephone number, work location, or Global Positioning System (GPS) data; addition or deletion of days of use for an already registered location; change in information about already registered devices; or updating of the expiration date of a license. Changes filed electronically or by facsimile are considered acceptable if the licensee receives telephone or facsimile confirmation that the NRC has received the submittal.

Exclusive Federal Jurisdiction. An area over which the Federal government exercises legal control without interference from the jurisdiction and administration of State law.

Filing. Filing is the submittal of documents to the NRC to report proposed activities in non-Agreement States, areas of exclusive Federal jurisdiction, or offshore waters. Filing of Form 241 will be considered complete when the documents are received and acknowledged by the NRC.

Initial Filing. Notification to the NRC on the first NRC Form 241 filed within a calendar year by an Agreement State licensee requesting reciprocity for activities conducted in non-Agreement States, areas of exclusive Federal jurisdiction, or offshore waters. This initial filing may be for multiple locations, clients, and unspecified dates of use until work activities are initiated. NRC Form 241 shall be filed at least 3 days before engaging in the proposed activity for the first time. The filing must include completed information on NRC Form 241, the appropriate fee, and one copy of the Agreement State license authorizing the proposed activity.

Non-Agreement State. Any State that is not an Agreement State.

Offshore Waters. That area of land and water, beyond Agreement States' Submerged Lands Act jurisdiction, on or above the U.S. Outer Continental Shelf.

Reciprocity. Commission recognition of applicable Agreement State licenses for work performed in areas of NRC jurisdiction. This term is also used in Agreement States with regard to Agreement State recognition of NRC licenses, as well as licenses from other Agreement States, for work performed within their jurisdiction.

Reciprocity Activities. Activities conducted by Agreement State licensees in non-Agreement States, areas of exclusive Federal jurisdiction, and offshore waters, under the general license provisions of Title 10 of the *Code of Federal Regulations* (10 CFR) 150.20, "Recognition of Agreement State Licenses."

1. PURPOSE OF REPORT

This report provides information important to Agreement State licensees on submitting NRC Form 241, "Report of Proposed Activities in Non-Agreement States, Areas of Exclusive Federal Jurisdiction, or Offshore Waters" (see Appendix A). For initial engagement in licensed activities in NRC jurisdiction during a calendar year, the NRC regional office where the Agreement State is located must receive Form 241 at least three (3) days before the Agreement State licensee engages in activities permitted under the General License established by Title 10 of the *Code of Federal Regulations* (10 CFR) 150.20, "Recognition of Agreement State Licenses." For changes to NRC Form 241, submitted changes should be complete and receive NRC review before the licensee implements the change. This report also contains the NRC's criteria for evaluating NRC Form 241. In addition, it provides NRC or Agreement State licensees seeking to conduct licensed activities in areas subject to the regulatory authority of an Agreement State with basic information on contacting the appropriate Agreement State Radiation Control Program Office to either obtain a specific license from that Agreement State or comply with the reciprocity requirements of that Agreement State. This information on contacting the appropriate Agreement State Radiation Control Program Office would also apply to Agreement State licensees seeking to conduct licensed activities within another Agreement State.

This document is intended to be used for all NRC byproduct, source, and special nuclear material licenses that are licensed in accordance with 10 CFR Parts 30, 34, 35, 39, 40, and 70; and for all Agreement State byproduct, source, and special nuclear material licenses that are licensed in accordance with the corresponding Agreement State regulations, provided that they include authorization for the use of licensed material at temporary job sites. (Note: The text of these and other NRC regulations may be found at http://www.nrc.gov/reading-rm/doc-collections/cfr/.)

It is necessary for a licensee to verify the jurisdictional status of the area of proposed activities to determine whether the NRC or an Agreement State has regulatory authority. Chapter 2 of this report provides information on determining the jurisdictional status of areas of proposed activities. Information about the specific requirements for particular uses of licensed materials may be found in the applicable volume of the NUREG-1556, "Consolidated Guidance about Materials Licenses," series, or in other appropriate NRC guidance documents.

This report also identifies general information that is needed to request changes to the information provided in an initial NRC Form 241 (see Appendix A). The information collection requirements in 10 CFR Parts 30, 34, 35, 39, 40, and 70 and in NRC Form 241 have been approved under Office of Management and Budget (OMB) Clearance Nos. 3150-0007, 3150-0009, 3150-0010, 3150-0013, 3150-0017, 03150-0020, and 3150-0130, respectively.

NRC Form 241 may not have sufficient space for applicants to provide full responses to Items 8 through 15 for additional clients, locations, and specific dates of scheduled work activities; as indicated on the form, the information pertinent to those items is to be provided on separate sheets of paper and submitted with the completed NRC Form 241. Chapter 8 of this report provides information for properly protecting proprietary information, such as a client list, that is submitted to the NRC along with NRC Form 241.

2. RECIPROCAL RECOGNITION OF SPECIFIC LICENSES

NRC regulations in 10 CFR 30.3, "Activities Requiring License," require that, "except…for persons exempt as provided for in this part and Part 150 of this chapter, no person shall manufacture, produce, transfer, receive, acquire, own, possess, or use byproduct material except as authorized in a specific or general license issued in accordance with the regulations in this chapter."

NRC and Agreement State specific licensees are responsible for determining, in advance, the jurisdictional status of the temporary job site where they plan to conduct licensed activities.

1556-049 ppt
050400

Figure 2.1 Reciprocity: Performing work in other jurisdictions is possible through reciprocal recognition of specific licenses.

In 10 CFR 150.20, the NRC provides that, under certain conditions, "any person who holds a specific license from an Agreement State, where the licensee maintains an office for directing the licensed activity and retaining radiation safety records, is granted a general license to conduct the same activity in—(i) Non-Agreement States; (ii) Areas of exclusive Federal jurisdiction within Agreement States; and (iii) Offshore waters."

Note that the requirements in 10 CFR150.20(b)(5) require that those operating under a general license are subject to NRC regulations as well as those terms and conditions of their Agreement State license.

Agreement State licensees who wish to conduct licensed activities in a non-Agreement State, in an area of exclusive Federal jurisdiction within an Agreement State, or in offshore waters should contact the NRC regional office for the Agreement State that issued their license. See Figure 2.2 for the appropriate addresses, telephone numbers, and a map of the Agreement States.

> Activities conducted under the general license granted by 10 CFR 150.20 are limited to 180 days in any calendar year. An exception to this limit is that activities in offshore waters may be conducted for an unlimited period of time. Throughout the year, Agreement State licensees should track work activities to ensure that they do not exceed the 180-day limit.

Agreement State licensees who propose to conduct licensed activities in NRC jurisdiction must either obtain a specific NRC license or file for reciprocity by submitting an NRC Form 241 with the appropriate NRC regional office in which the Agreement state that issued the license is located. An NRC Form 241 shall be filed at least three (3) days before engaging in each activity for the first time in a calendar year. An NRC-signed copy of NRC Form 241 will be returned to the licensee to acknowledge its approval.

If a submittal cannot be filed 3 days before the proposed activity because of an unusual circumstance such as an emergency the Regional Administrator may waive the 3-day time requirement, provided that the remaining provisions of 10 CFR 150.20(b)(1) are satisfied. Emergency requests should be directed to the appropriate regional office by telephone before initiation of the activity, and will be reviewed on a case by case basis.

Chapter 7 of this report provides specific instructions on completing NRC Form 241. Agreement State licensees conducting licensed activities in NRC jurisdiction must comply with the terms of their Agreement State license and with all applicable NRC rules and regulations.

The intent of 10 CFR 150.20 is to allow Agreement State licensees to work at temporary job sites in NRC jurisdiction for a limited period of time without having to obtain a specific NRC license. If a company has more than one Agreement State license, it must submit a separate NRC Form 241 for work conducted under each Agreement State license used during the calendar year.

If the company has more than one Agreement State license and proposes to perform the same type of licensed activity under another separate license, it cannot exceed the 180-day limit under reciprocity for that work activity. For example, a company authorized to conduct radiography under both a Texas license and a Mississippi license will have its radiography activities in NRC jurisdiction limited to 180 days in a calendar year.

However, a legal entity holding two or more Agreement State licenses for different activities may file one NRC Form 241 for each type of activity in any calendar year. For example, a company authorized to conduct radiography under a New Mexico license and gauge work under a Colorado license may file a separate NRC Form 241 for each activity, allowing 180 days for radiography and 180 days for gauge work in NRC jurisdiction in a calendar year.

2.1 Enforcement

> Failure to file NRC Form 241 or obtain approval from the Regional Administrator can result in enforcement action, which may include civil penalties.

Agreement State licensees may be subject to NRC enforcement action should they fail to file for reciprocity or obtain a specific NRC license before working in non-Agreement States, areas of exclusive Federal jurisdiction within Agreement States, or in offshore waters. In addition, Agreement State licensees operating under reciprocity within NRC jurisdiction are subject to NRC inspection and to enforcement actions when violations are found during an NRC inspection.

For information on NRC inspection, investigation, enforcement, and other compliance programs, see the current version of the NRC Enforcement Policy and Guidance and the appropriate NRC Inspection Manual chapters. The current version of the NRC Enforcement Policy and Guidance is available electronically at http://www.nrc.gov/about-nrc/regulatory/enforcement.html.

2.2 Reporting Requirements

In accordance with 10 CFR 20.2201, "Reports of theft or loss of license material," licensees are required to report the theft or loss or radioactive material. In addition, licensees in accordance with 10 CFR 30.50, "Reporting Requirements," are required to report instances that could lead to exposures to radiation or radioactive materials. This includes failures of safety components, unplanned contamination, etc.

For immediate notifications, licensees can contact the NRC Operations Center at (301) 816-5100.

2.3 Federal Agency Activities

Federal agencies licensed by the NRC are not subject to Agreement State radiation control regulations. Thus, Federal licensees who are authorized to work at temporary job sites under their NRC licenses may do so anywhere in the United States and its territories.

2.4 Manufacturer, Installer, and Servicer Activities

NRC regulations in 10 CFR 31.6, "General License To Install Devices Generally Licensed in § 31.5," refer to the installation of generally licensed devices under 10 CFR 31.5, "Certain Detecting, Measuring, Gauging, or Controlling Devices and Certain Devices for Producing Light or an Ionized Atmosphere," and other authorized activities. The manufacturer, installer, or servicer of a device who holds a specific license as described in §31.5 from the NRC or an Agreement State, and meets the requirements in §31.6, may perform those functions anywhere in NRC jurisdiction without filing for reciprocity, including in offshore waters as defined in 10 CFR 150.20.

2.5 Activities in Agreement States

NRC licensees who wish to conduct licensed activities in an Agreement State and Agreement State licensees who wish to conduct licensed activities in an Agreement State other than their own should contact the appropriate Agreement State's Radiation Control Program Office for information about that State's regulations and filing process.

Reciprocity filing relates to NRC licensees, particularly radiographers, well loggers, and portable gauge users, who are authorized to use radioactive material at temporary job sites. However, this authorization does not permit operations in Agreement State jurisdiction (except for licensees who are Federal agencies). In these circumstances, NRC licensees and Agreement State licensees are subject to the appropriate Agreement State's regulations.

Agreement States have reciprocity provisions that permit NRC licensees to perform work using NRC-licensed radioactive materials in their States to ensure compliance with the Agreement State's reciprocity requirements, licensees are advised to request authorization from the appropriate regulatory authority well in advance of the scheduled use of licensed material. A list of Agreement State agency contacts is available at the Office of Federal and State Materials and Environmental Management Programs (FSME) public Web site at http://nrc-stp.ornl.gov/asdirectory.html.

Table 2.1 Do I Need To File for Reciprocity?[1]

Reciprocity applicants may use the Table below to determine whether reciprocity filing is needed **once the regulatory authority over the proposed work location has been determined**. The table may not cover all jurisdictional situations. Contact the appropriate Regional office if there are questions about work activities at a particular area.

License Issued by	Proposed Work Location	Regulatory Authority	Reciprocity Needed?
NRC to a Federal agency	Agreement State or non-Agreement State	NRC	No
NRC to licensee other than a Federal agency	In non-Agreement State, U.S. territory, possession, or in offshore Federal waters	NRC	No
NRC to licensee other than a Federal agency	Agreement State at Federally controlled site subject to exclusive Federal jurisdiction (e.g. VA hospital)	NRC	No
NRC to licensee other than a Federal agency	Agreement State at Federally controlled site **not** subject to exclusive Federal jurisdiction (e.g., private hospital)	Agreement State	Yes—file with Agreement State
Agreement State	Non-Agreement State or other areas under NRC jurisdiction	NRC	Yes—file with NRC
Agreement State	Exclusive Federal jurisdiction—within an Agreement State	NRC	Yes—file with NRC
Agreement State	Other Agreement State at Federally controlled site not subject to exclusive Federal jurisdiction	Other Agreement State	Yes—file with Agreement State having regulatory authority

---

[1]   For additional jurisdictional guidance refer to the FSME procedures in the State Agreement (SA) series, SA-500, "Jurisdiction Determination," which is available at http://nrc-stp.ornl.gov/

License Issued by	Proposed Work Location	Regulatory Authority	Reciprocity Needed?
Agreement State	Other Agreement State	Other Agreement State	Yes—file with Other Agreement State
Agreement State	Federally recognized Indian Tribe reservation or Tribal areas of exclusive federal jurisdiction	NRC	Yes—file with NRC
Agreement State	In Agreement State conducting industrial radiography at a Part 50 or 52 rector site, including construction, pre-operational and operational phases	NRC	Yes—file with NRC
NRC	In Agreement State conducting industrial radiography at a Part 50 or 52 rector site, including construction, pre-operational and operational phases	NRC	No

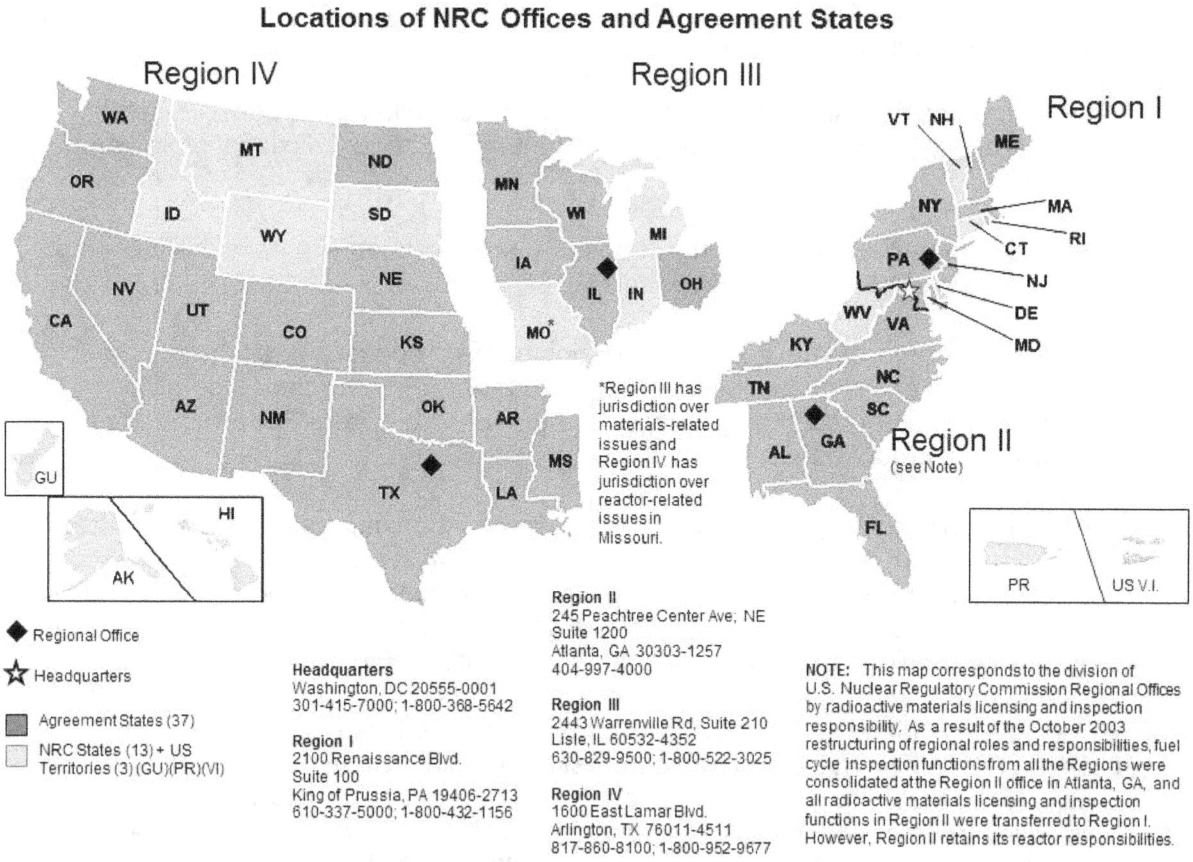

Locations of NRC Offices and Agreement States

Region IV Region III Region I

*Region III has jurisdiction over materials-related issues and Region IV has jurisdiction over reactor-related issues in Missouri.

Region II (see Note)

◆ Regional Office

☆ Headquarters

▓ Agreement States (37)

▢ NRC States (13) + US Territories (3) (GU)(PR)(VI)

Headquarters
Washington, DC 20555-0001
301-415-7000; 1-800-368-5642

Region I
2100 Renaissance Blvd.
Suite 100
King of Prussia, PA 19406-2713
610-337-5000; 1-800-432-1156

Region II
245 Peachtree Center Ave; NE
Suite 1200
Atlanta, GA 30303-1257
404-997-4000

Region III
2443 Warrenville Rd, Suite 210
Lisle, IL 60532-4352
630-829-9500; 1-800-522-3025

Region IV
1600 East Lamar Blvd.
Arlington, TX 76011-4511
817-860-8100; 1-800-952-9677

NOTE: This map corresponds to the division of U.S. Nuclear Regulatory Commission Regional Offices by radioactive materials licensing and inspection responsibility. As a result of the October 2003 restructuring of regional roles and responsibilities, fuel cycle inspection functions from all the Regions were consolidated at the Region II office in Atlanta, GA, and all radioactive materials licensing and inspection functions in Region II were transferred to Region I. However, Region II retains its reactor responsibilities.

Figure 2.2 U.S. map: locations of NRC offices and Agreement States

2.6 Exclusive Federal Jurisdiction

An area of exclusive Federal jurisdiction is an area over which the Federal Government exercises legal control without interference from the jurisdiction and administration of State law. This means that, in such areas, the Federal Government has sole jurisdiction for both civil (e.g., regulatory) and criminal matters. This sole jurisdiction includes issues involving health and environmental protection, such as the regulation of radioactive material use. Federal ownership of land does not necessarily mean that licensees are subject to NRC regulatory control when working on that site. Whether a Federal enclave is an area of exclusive Federal jurisdiction must be determined case by case because the status of such land is subject to change. Many, but not all, of these areas are on military bases where, for purposes of national security, the Federal Government maintains exclusive control.

Agreement State licensees cannot conduct licensed operations in areas of exclusive Federal jurisdiction without either (1) filing NRC Form 241 for reciprocity in accordance with 10 CFR 150.20(b) or (2) obtaining a specific NRC license. Information about the specific NRC requirements for particular uses of licensed materials may be found in the applicable volume of the NUREG-1556 series or other appropriate guidance documents.

The NRC cannot effectively maintain accurate information about the jurisdictional status of Federal lands or facilities throughout the country because of the large number of sites and the fact that their status is constantly changing.

If you intend to conduct licensed activities at a Federally controlled site (e.g., a Federally controlled site in an Agreement State), you should determine the jurisdictional status of the site.

If you are uncertain about the jurisdictional status of a proposed job site, take the following steps (The steps can also be found in Appendix B):

(1)   Obtain specific information about the location of the proposed job site (e.g., street address, range/township, building or hangar number, distance from a specific intersection, or other identifying details) and the **identity of the Federal agency** controlling the proposed job site.

(2)   Call the Federal agency's local contact (e.g., contract officer, base environmental health officer, judge advocate general district office staff, regional office staff, regional counsel) and request a written statement or supporting information about the jurisdictional status of the proposed job site. Document the name and title of the person at the Federal agency who provided the determination and the date that it was provided.

(3)   If the job site is identified as falling under "Exclusive Federal Jurisdiction" or other NRC licensed facility that has a controlled access/protected area, and you are an Agreement State licensee, submit to the appropriate NRC Regional office shown above your notification of proposed work (NRC Form 241) and, if available, a copy of the statement of jurisdiction from the agency. In lieu of submitting an NRC Form 241, Agreement State licensees may apply for a specific NRC license to operate in areas under NRC jurisdiction. If you are an NRC licensee, no action is required.

(4)   If the job site is identified as other than "Exclusive Federal Jurisdiction" you should contact the Agreement State within which the facility exists.

(5)   If the job site is on Federally recognized Tribal reservation or other Tribal areas of exclusive federal jurisdiction I in an Agreement State, you should contact the appropriate NRC regional office as specified in 10 CFR 30.6, "Communications" to determine whether NRC or the Agreement State has regulatory authority over these areas. In Agreement States, non-Federally-recognized Tribal lands fall under the Agreement State regulatory authority.

An Agreement State licensee found to be conducting licensed activities in an area of exclusive Federal jurisdiction without an NRC license or without filing for reciprocity under 10 CFR 150.20 is potentially subject to escalated enforcement action, including civil penalties and orders. However, the NRC will typically not take enforcement action against an Agreement State licensee for such violations if the licensee has evidence that it received a determination from the Federal agency that the area is not under exclusive Federal jurisdiction. This evidence may be a written statement from the Federal agency that provided the determination and the date that it was provided, or a written record of the name and title of the person at the Federal agency who provided the determination and the date that it was provided.

For information on NRC inspection, investigation, enforcement, and other compliance programs, see the current version of the NRC Enforcement Policy and Guidance, and the appropriate NRC Inspection Manual chapters. The current version of the NRC Enforcement Policy and Guidance is available electronically at http://www.nrc.gov/about-nrc/regulatory/enforcement.html.

2.7 Offshore Waters and the Great Lakes

Offshore waters are defined as that area of land and water, beyond Agreement States' Submerged Lands Act jurisdiction, on or above the U.S. Outer Continental Shelf. For most states, the NRC maintains jurisdiction over licensed activities conducted in offshore waters beyond 3 nautical miles off the coastline. The NRC has jurisdiction beyond 9 nautical miles off the coastline for the State of Texas and Florida's west coast. Effective April 26, 2000, the NRC resumed jurisdiction over activities in offshore waters adjacent to the shoreline of the State of Louisiana.

NRC regulation 10 CFR 34.41(c) requires that a licensee may conduct lay-barge, offshore platform or underwater radiography only if procedures have been approved by the NRC or by an agreement state. Licensees need to provide operating and emergency procedures for lay-barge, offshore platform or underwater radiography that addresses the following safety aspects:

- Procedures, as required to comply with 10 CFR 34.45(a), unique to offshore platform radiography, to minimize the potential for a source disconnect.

- Emergency procedures, as required by 10 CFR 34.45(a), unique to offshore platform radiography, to address unintentional source disconnects and recovery. These procedures should address minimizing radiation exposure for all personnel on the offshore platform during a disconnect event, Including considerations such as staff training, availability of portable shielding, survey instruments, dosimetry, and other equipment.

- Operating procedures and practices, unique to offshore platform radiography to ensure compliance with 10 CFR 34.41 requirements for personnel to be present during radiography and 10 CFR 34.46 requirements for supervision of radiography assistants.

- Operating procedures and practices, as required to comply with 10 CFR 34.45(a) to ensure that radiography personnel have the authority necessary to protect personnel on an offshore platform from radiation doses in excess of regulatory limits. These should include provisions to ensure that radiographic operations are conducted only in those locations shielded for such operations (if those enclosures are relied upon to control dose) and that radiographic personnel can direct personnel as necessary for radiation safety both during normal operations and in the event of a disconnect or other emergency.

Contact your state licensing agency and/or the NRC if questions arise about the extent of jurisdiction in offshore waters. These specific questions will be decided on a case-by-case basis.

In situations involving the determination of regulatory jurisdiction over submerged lands in the Great Lakes for the purposes of filing reciprocity, licenses should determine whether their job-site is located south of the international boundary line running through the Great Lakes.

The licensee would then determine which state has regulatory authority over the location of their job-site. If the location is within the designated boundaries of an Agreement State waters, they should contact the appropriate Agreement State for reciprocity. If the waters are subject to a non-Agreement State regulatory authority, such as Michigan, the licensee should contact the NRC for reciprocity.

In 10 CFR 150.20(b)(4), the NRC states, in part, that the general license granted in 10 CFR 150.20(a) concerning activities in offshore waters authorizes any person who holds a specific license from an Agreement State to engage in the activities authorized for an unlimited period of time. In other words, activities conducted under reciprocity in offshore waters are not limited to 180 days in any calendar year. However, in accordance with 10 CFR 150.20(b)(1), an Agreement State licensee is required to file an initial NRC Form 241 for each calendar year in which it conducts activities in offshore waters.

2.8 Nuclear Reactor Facilities

The possession and use of radioactive materials at a reactor facility would be covered under the NRC Part 50 or 52 license when the materials are within the protected or controlled access areas of the NRC licensed facility.

Note: If the reactor facility is located within an Agreement or non-Agreement State, the NRC is the regulatory authority on site, and an Agreement State licensee should file for reciprocity by submitting NRC Form 241, as required by 10 CFR 150.20, or obtain a specific license issued by the NRC under 10 CFR Part 34, "Licenses for Industrial Radiography and Radiation Safety Requirements for Industrial Radiographic Operations."

For other NRC licensed facilities in Agreement States, the licensee should contact the NRC Office of General Counsel to determine whether reciprocity should be filed with NRC or the Agreement State.

For more information about how to determine exclusive federal jurisdiction, see Appendix B to this volume or review the NRC's State Agreement (SA) Series, SA-500, "Jurisdiction Determination," at http://nrc-stp.ornl.gov. Questions about exclusive federal jurisdiction determination should be referred to the appropriate Regional Office.

2.9 Tribal Lands

On Tribal land, as implemented in 10 CFR Part 150, "Exemptions and Continued Regulatory Authority in Agreement States and in Offshore Waters under Section 274," the NRC does not transfer regulatory authority to the States for Federally recognized Tribe reservations or Tribal areas of exclusive federal jurisdiction. Consequently, absent NRC written approval, a State may not regulate activities on a Federally recognized Tribe reservation or Tribal areas of exclusive federal jurisdiction. Agreement States have regulatory authority over non-Federally recognized Tribal lands.

Licensees who propose to conduct licensed activities on a federally recognized Tribe reservation or Tribal areas of exclusive federal jurisdiction should contact the appropriate NRC regional office. The NRC will enter into a dialogue with the involved Stakeholders (Agreement State, Native American representatives, licensee or applicant for approval of the proposed activity under the Atomic Energy Act of 1954, as amended (AEA)) concerning the jurisdictional question for the particular activity.

Tribal Governments or Tribal members seeking a radioactive material license should contact the appropriate NRC regional office.

Note: A federally recognized Tribe is an American Indian or Alaska Native tribal entity that is recognized as having a government-to-government relationship with the United States, with the responsibilities, powers, limitations, and obligations attached to that designation, and is eligible for funding and services from the Bureau of Indian Affairs. State only recognized Tribes are those that have not been recognized by the federal government. The status of the Tribe and tribal land will determine whether the NRC or Agreement State has jurisdiction authority over the activities in a specific area.

For more information about how to determine exclusive federal jurisdiction, see Appendix B to this volume or review the NRC's State Agreement (SA) Series, SA-500, "Jurisdiction Determination," at http://nrc-stp.ornl.gov. Questions about exclusive federal jurisdiction determination should be referred to the appropriate Regional Office.

2.10 Commercial Launch Operations

Commercial launches are carried out by the Office of the Associate Administrator for Commercial Space Transportation of the Federal Aviation Administration (FAA). The FAA's jurisdiction over commercial space projects begins with the arrival of a launch vehicle at any launch site located in the United States, and its jurisdiction extends to accidental payload reentries and in-orbit activities arising out of a launch.

Currently, commercial launches take place from Federal launch ranges operated by the U.S. Department of Defense and the National Aeronautics and Space Administration. Launch operators bring vehicles for launch to Federal ranges, such as Cape Canaveral Air Station, Vandenberg Air Force Base, White Sands Missile Range, and Wallops Flight Facility. Licensees who propose to conduct licensed activities as part of a space launch should contact the launch operators about jurisdiction at the launch range.

To the extent that a commercial space venture would involve the possession and/or use of nuclear materials *before* any space vehicle arrives at the launch site, the NRC would exercise jurisdiction within non-Agreement States and in areas of exclusive Federal jurisdiction within an Agreement State.

2.11 Activities in Antarctica

Antarctica is defined as that area of the world south of 60 degrees south latitude, as described in the International Antarctic Treaty (1959) and the Antarctic Conservation Act (16 U.S.C. 2401 et seq.).

Presidential Memorandum 6646, dated February 5, 1982, designates the National Science Foundation (NSF) as the executive Federal manager for the United States Antarctic Program, assigning the entire management responsibility to NSF. This includes activities involving the use of source, byproduct, and special nuclear material.

Within its regulatory framework, NSF has adopted as policy the radiation protection standards for byproduct, source, and special nuclear material set forth in the NRC regulations in 10 CFR Part 20, "Standards for Protection against Radiation," and U.S. Department of Transportation (DOT) regulations for packaging and transport of such radioactive materials. NSF ensures that the disposal of radioactive waste generated in Antarctica (but returned to the United States for disposal) is consistent with NRC, U.S. Environmental Protection Agency, and Agreement State requirements.

Because of similar responsibilities, the NRC and NSF issued a Memorandum of Understanding (MOU) on September 3, 1999. The purpose of this MOU is to clarify the responsibilities of NSF and the NRC with respect to the safe use of radioactive materials in Antarctica, in order to avoid any duplication of effort and to ensure efficient management of those materials. The MOU provides that the NRC will not exercise any authority under the AEA to regulate the use of byproduct, source, and special nuclear material in Antarctica and recognizes NSF's regulatory oversight.

Work in Antarctica does not require reciprocity filing. NRC and Agreement State licensees are encouraged to implement programs to ensure the safe use of radioactive materials when performing work activities in Antarctica under NSF jurisdiction. The NRC or Agreement States, as appropriate, will resume regulatory authority over any byproduct, source, or special nuclear material that is regulated by NSF in Antarctica if and when such material reenters the United States or its territories.

Information about activities in Antarctica may be found on the NSF's Web site at http://www.nsf.gov. As an alternative, contact the Director, Office of Polar Programs, National Science Foundation, 4201 Wilson Boulevard, Arlington, VA 22230.

2.12 Activities in U.S. Territories

Other areas over which the NRC maintains jurisdiction are the U.S. Virgin Islands, Puerto Rico, and other U.S. territories, such as American Samoa, Johnston Atoll, and Guam. Agreement State licensees who plan work activities in U.S. territories must file NRC Form 241 to the appropriate NRC regional office.

2.13 Activities in Agreement States

If the work location is an Agreement State and is not an area of exclusive Federal jurisdiction nor within the protected or controlled access area of NRC licensed facility, these activities are regulated by the Agreement State Radiation Control Program Office, not by the NRC. Each Agreement State has its own regulations that allow reciprocity activities within its State. The NRC and licensees from other Agreement States should contact the Agreement State Radiation Control Program Office in which the work is located as noted in Section 2.4

3. MANAGEMENT RESPONSIBILITY

The NRC recognizes that effective radiation safety program management is vital to achieving safe, secure, and compliant operations. The NRC believes that consistent compliance with its regulations provides reasonable assurance that licensed activities will be conducted safely and that effective management will result in increased safety, security, and compliance.

> "Management" as used in this volume refers to the processes for conduct and control a radiation safety program and to the individuals who are responsible for those processes and who have *authority to provide necessary resources* to achieve regulatory compliance.

3.1 Commitments and Responsibilities

It is the Agreement State licensee's obligation to keep the information submitted in support of the general license established by 10 CFR 150.20 current. The licensee must submit changes that affect any of the information provided in the initial NRC Form 241.

The general license in 10 CFR 150.20 is established on a calendar-year basis. Therefore, an Agreement State licensee proposing to conduct licensed activities in NRC jurisdiction must file an initial NRC Form 241 before engaging in such activities for the first time in *each* calendar year. Licensed activities conducted in NRC jurisdiction, except for those conducted in offshore waters, are limited to 180 days in any calendar year.

Generally, Agreement State licensee management has a responsibility for all aspects of the radiation safety program, including, but not limited to, the following:

- Radiation safety, security, and control of radioactive materials, and compliance with regulations;

- Completeness and accuracy of the radiation safety records and all information provided to the NRC (10 CFR 30.9, 10 CFR 40.9, 10 CFR 70.9, "Completeness and accuracy of information");

- Knowledge of the contents of the current Agreement State license and NRC Form 241;

- Compliance with current NRC and DOT regulations and the licensee's operating and emergency procedures;

- Commitment to provide adequate resources (including space, equipment, personnel, time, and, if needed, contractors) to the radiation protection program to ensure that the public and workers are protected from radiation hazards and that compliance with regulations is maintained;

- Selection and assignment of a qualified individual to serve as the radiation safety officer for licensed activities; and

- Commitment to ensure that workers have had adequate training.

The NRC expects Agreement State licensee management to conduct its licensed activities within NRC jurisdiction in compliance with all terms and conditions of the specific license issued by the Agreement State, except such terms and conditions as are contrary to the requirements of 10 CFR 150.20. The general licenses provided in 10 CFR 150.20 are subject to all of the provisions of the Atomic Energy Act of 1954, now or hereafter in effect, and to all applicable rules, regulations, and orders of the Commission. Therefore, licensees are expected to notify the NRC in a timely manner of a change that affects the information provided in the initial NRC Form 241, allowing the NRC adequate time to inspect the activity and carry out its statutory mandate of ensuring that licensed materials are adequately safeguarded and that public health and safety are protected.

For information on NRC inspection, investigation, enforcement, and other compliance programs, see the current version of the NRC's Enforcement Policy and Inspection Procedures available on the NRC's website.

When working under reciprocity, Agreement State licensees must comply with current NRC requirements. Licensees need to be aware that NRC regulations may differ in some respects from Agreement State regulations. In addition, NRC regulations may change to include new requirements before the Agreement States amend their regulations to include similar requirements.

3.2 Safety Culture

Individuals and organizations performing regulated activities are expected to establish and maintain a positive safety culture commensurate with the safety and security significance of their activities and the nature and complexity of their organizations and functions. This applies to all licensees, certificate holders, permit holders, authorization holders, holders of quality assurance program approvals, vendors and suppliers of safety-related components, and applicants for a license, certificate, permit, authorization, or quality assurance program approval, subject to NRC authority.

"Nuclear safety culture" is defined in the NRC's safety culture policy statement (76 FR 34773; June 14, 2011) as *the core values and behaviors resulting from a collective commitment by leaders and individuals to emphasize safety over competing goals to ensure protection of people and the environment.* Individuals and organizations performing regulated activities bear the primary responsibility for safely handling and securing these materials. Experience has shown that certain personal and organizational traits are present in a positive safety culture. A trait, in this case, is a pattern of thinking, feeling, and behaving that emphasizes safety, particularly in goal conflict situations (e.g., production versus safety, schedule versus safety, and cost of the effort versus safety). Refer to Table 3.1 for the traits of a positive safety culture from NRC's safety culture policy statement.

Organizations should ensure that personnel in the safety and security sectors have an appreciation for the importance of each, emphasizing the need for integration and balance to achieve both safety and security in their activities. Safety and security activities are closely intertwined. While many safety and security activities complement each other, there may be instances in which safety and security interests create competing goals. It is important that consideration of these activities be integrated so as not to diminish or adversely affect either; thus, mechanisms should be established to identify and resolve these differences. A safety culture that accomplishes this would include all nuclear safety and security issues associated with NRC-regulated activities.

The NRC, as the regulatory agency with an independent oversight role, reviews the performance of individuals and organizations to determine compliance with requirements and commitments through its existing inspection and assessment processes. However, NRC's safety culture policy statement and traits are not incorporated into the regulations.
Many of the safety culture traits may be inherent to an organization's existing radiation safety practices and programs. For instance, working under reciprocity would generally include transportation of radioactive material. Licensee operating controls and procedures for transporting licensed material (e.g. package compliance, external radiation level measurements, package security) developed to meet the transportation requirements may correspond with the safety culture traits specific in Table 3.1 as "Work Processes" (the process of planning and controlling work activities is implemented so that safety is maintained).

Refer to Appendix M for the NRC's safety culture policy statement. More information on NRC activities relating to safety culture can be found at:
http://www.nrc.gov/about-nrc/regulatory/enforcement/safety-culture.html.

Table 3.1 Traits of a Positive Safety Culture

Leadership Safety Values and Actions	Problem Identification and Resolution	Personal Accountability
Leaders demonstrate a commitment to safety in their decisions and behaviors	Issues potentially impacting safety are promptly identified, fully evaluated, and promptly addressed and corrected commensurate with their significance	All individuals take personal responsibility for safety
Work Processes	**Continuous Learning**	**Environment for Raising Concerns**
The process of planning and controlling work activities is implemented so that safety is maintained	Opportunities to learn about ways to ensure safety are sought out and implemented	A safety conscious work environment is maintained where personnel feel free to raise safety concerns without fear of retaliation, intimidation, harassment or discrimination
Effective Safety Communications	**Respectful Work Environment**	**Questioning Attitude**
Communications maintain a focus on safety	Trust and respect permeate the organization	Individuals avoid complacency and continuously challenge existing conditions and activities in order to identify discrepancies that might result in error or inappropriate action

4. APPLICABLE REGULATIONS

It is the applicant or licensee's responsibility to obtain and have available up-to-date copies of applicable regulations, to read and understand the requirements of each of these regulations, and to comply with each applicable regulation. The following parts of 10 CFR contain regulations applicable to licensing of byproduct, source, and special nuclear materials. Some of these parts are specific to one type of license, while others are general and will apply to many if not all licensees.

The current versions of these parts can be found under the "Basic References" link at the NRC's online library at http://www.nrc.gov/reading-rm.html; if viewing in a browser, the following list includes a direct link to the rules:

- 10 CFR Part 2, "Agency Rules of Practice and Procedure"

- 10 CFR Part 19, "Notices, Instructions and Reports to Workers: Inspection and Investigations"

- 10 CFR Part 20, "Standards for Protection Against Radiation"

- 10 CFR Part 21, "Reporting of Defects and Noncompliance"

- 10 CFR Part 30, "Rules of General Applicability to Domestic Licensing of Byproduct Material"

- 10 CFR Part 31, "General Domestic Licenses for Byproduct Material"

- 10 CFR Part 34, "Licenses for Industrial Radiography and Radiation Safety Requirements for Industrial Radiographic Operations"

- 10 CFR Part 35, "Medical Use of Byproduct Material"

- 10 CFR Part 37, "Physical Protection of Category 1 and 2 Quantities of Radioactive Material"

- 10 CFR Part 39, "Licenses and Radiation Safety Requirements for Well Logging"

- 10 CFR Part 40, "Domestic Licensing of Source Material"

- 10 CFR Part 50, "Domestic Licensing of Production and Utilization Facilities"

- 10 CFR Part 51, "Environmental Protection Regulations for Domestic Licensing and Related Regulatory Functions"

- 10 CFR Part 70, "Domestic Licensing of Special Nuclear Material"

- 10 CFR Part 71, "Packaging and Transportation of Radioactive Material"

- 10 CFR Part 110, "Export and Import of Nuclear Equipment and Material"

- 10 CFR Part 150, "Exemptions and Continued Regulatory Authority in Agreement States and in Offshore Waters Under Section 274"

- 10 CFR Part 170, "Fees for Facilities, Materials, Import and Export Licenses, and Other Regulatory Services Under the Atomic Energy Act of 1954, as Amended"

- 10 CFR Part 171, "Annual Fees for Reactor Licenses and Fuel Cycle Licenses and Materials Licenses, Including Holders of Certificates of Compliance, Registrations, and Quality Assurance Program Approvals and Government Agencies Licensed by the NRC"

Copies of the above documents may be obtained by calling the Government Printing Office order desk toll-free at (866) 512-8600, or in Washington, DC, at (202) 512-1800, or online at http://bookstore.gpo.gov.

A single copy of the above documents may be requested from the NRC's regional offices (see Figure 2.2 for addresses and telephone numbers). In addition, 10 CFR Parts 1 through 199 can be found on the NRC's Web site at http://www.nrc.gov/reading-rm/doc-collections/cfr under Regulations (10 CFR).

NRC regulations and amendments can also be accessed from the "NRC Library" link on the NRC's public Web site at http://www.nrc.gov. The NRC and all other Federal agencies publish amendments to their regulations in the *Federal Register*.

5. HOW TO FILE

5.1 Paper Application

Agreement State licensees requesting reciprocity should do the following:

- Complete NRC Form 241 (see Appendix A) Items 1 through 20.

- If additional space is needed to complete NRC Form 241, provide information in Items 8 through 18 on supplementary pages.

- For each separate sheet that is submitted with NRC Form 241, identify and key it to the item number on the form or the topic to which it refers.

- Submit all supplementary pages on 8-1/2 x 11-inch paper.

- At least 3 days before beginning activities, file an original NRC Form 241, one copy of the licensee's current Agreement State license, and the appropriate fee with the appropriate NRC regional office where the Agreement State licensee is located, not where they intend to perform work.

- Retain one copy of NRC Form 241 for future reference.

5.2 Where to File

Agreement State licensees who wish to conduct licensed activities in a non-Agreement State, in an area of exclusive Federal jurisdiction within an Agreement State, or in offshore waters should contact the NRC regional office for the Agreement State that issued their license. See Figure 2.2 for the appropriate addresses, telephone numbers, and a map of the Agreement States.

5.3 Acceptable Means of Initial Filing

In addition to direct mail, initial filing of NRC Form 241 by facsimile or e-mail is acceptable, provided that:

(1) the facsimile or e-mail contains, in addition to the completed NRC Form 241, a copy of the current Agreement State license and a copy of the check or credit card application that will be mailed to meet fee requirements, if appropriate,

(2) the Agreement State licensee confirms that the NRC has received the electronic submittal (confirmation of receipt by the NRC may be made by telephoning the NRC office to which the submittal was sent), and

(3) the NRC receives, within 3 days, NRC Form 241, one copy of the Agreement State license, and the appropriate fee, if applicable. Note that weekends and holidays may require additional processing time by the Regional office.

Changes to NRC Form 241 should be submitted to the NRC Regional Office either by facsimile or by e-mail to the electronic mailbox established by each office, if available.

Note: Avoid submitting proprietary information and personally identifiable information (such as home address, home telephone number, social security number, and date of birth) unless it is absolutely necessary or requested by the NRC. If it is necessary to submit proprietary information, follow the procedure in 10 CFR 2.390, "Public Inspections, Exemptions, Requests for Withholding." Failure to follow this procedure could result in disclosure of the proprietary information to the public or substantial delays in processing the form. Additional information on proprietary information is provided in Chapter 8, "Identifying and Protecting Sensitive Information," of this report.

6. RECIPROCITY FEES

Reciprocity fees are based on work activities within a calendar year. Submission of an initial NRC Form 241 must be accompanied by the appropriate fee. Agreement State licensees should refer to Category 16 of 10 CFR 170.31, "Schedule of Fees for Materials Licenses and Other Regulatory Services, Including Inspections, and Import and Export Licenses," to determine the amount of the initial fee. No fee is required for changes.

Fee payments are to be made payable to the U.S. Nuclear Regulatory Commission. The payments are to be made in U.S. funds by check, draft, money order, or credit card. NRC Form 629, "Authorization for Payment by Credit Card," is found in Appendix C. Specific instructions for making credit card payments may be obtained by contacting the Division of the Controller, Accounts Receivable/Payable Branch at **(301) 415-6097**. Note that credit card fees are processed by the U.S. Department of the Interior's National Business Center, the NRC's collection service provider.

The NRC will not process initial or revised NRC Form 241s before fee receipt. However, if the electronic or facsimile method is used to file for reciprocity, the NRC will process Form 241 as long as the submittal includes a copy of the check or credit card application that will be mailed to meet the applicable fee requirement. Once technical review has begun, no fees will be refunded; fees will be charged regardless of the NRC's disposition of an NRC Form 241 or the applicant's withdrawal of an NRC Form 241. The exception is a case in which it is determined that the proposed work location is *not* in NRC jurisdiction and therefore filing of NRC Form 241 is not warranted. Agreement State licensees should consult 10 CFR 170.11, "Exemptions," for additional information on exemptions from reciprocity fees.

Direct all questions about NRC fees to the Office of the Chief Financial Officer at NRC Headquarters in Rockville, MD, or at http://www.nrc.gov/about-nrc/organization/ocfofuncdesc.html.

7. CONTENTS OF NRC FORM 241

Agreement State licensees are required to complete NRC Form 241 identifying proposed activities in non-Agreement States, in areas of exclusive Federal jurisdiction within Agreement States, or in offshore waters. Licensees should file the completed NRC Form 241 with the NRC regional office for the Agreement State that issued their licenses.

In completing NRC Form 241, the Agreement State licensee must provide sufficient information to enable the NRC to plan and conduct inspections. If NRC Form 241 contains omissions or errors, the NRC staff will first try to resolve them by telephone with the applicant. If the discrepancies can be resolved by telephone, the NRC staff will mark the form with the corrections and continue processing the form. The NRC staff will sign the form and return a copy to the licensee once its review is completed.

If the deficiencies cannot be resolved by telephone, the NRC staff will either verbally request that applicant provide the deficient information or send a letter requesting the necessary information, identifying to the applicant the deficiencies and informing the licensee that the NRC will continue its review on receipt of the requested information. The Agreement State licensee will also be informed that work is not to be started in areas of exclusive Federal jurisdiction, non-Agreement States, or offshore waters until the NRC receives the required information and completes its review.

Agreement State licensees that do not meet the requirements of 10 CFR 150.20 will be informed that they do not qualify for the general license and may not conduct licensed activities in NRC jurisdiction. <u>Agreement State licensees that perform work in NRC jurisdiction before filing NRC Form 241 may be subject to NRC enforcement action.</u>

7.1 Item 1: Name of Licensee

List the licensee's name as identified on the Agreement State license (i.e., the person or firm proposing the activities).

7.2 Item 2: Type of Report

When submitting NRC Form 241, Agreement State licensees are required to mark the appropriate box: "initial" or "change."

Initial Filing

Agreement State licensees seeking to conduct activities under reciprocity in areas of exclusive Federal jurisdiction, non-Agreement States, or in offshore waters *for the first time in a calendar year* are required to submit NRC Form 241.

Changes

Changes provide information that modifies items approved in the initial NRC Form 241 filing. These may include changes to specific locations of work sites, additional work locations, changes to the radioactive material or work activities, changes in contacts or telephone numbers, or additional or deleted dates of work from the initial NRC Form 241. Changes should be provided in advance and on a timely basis to allow regional review and approval

7.3   Item 3: Address of Licensee

Provide the mailing address where correspondence should be sent. Use the mailing address specifically identified on the Agreement State license. Notify the NRC of any changes in mailing address on NRC Form 241.

7.4   Item 4: Licensee Contact and Title

Identify the individual who can answer questions about NRC Form 241. This is typically the radiation safety officer, unless the licensee has named a different person as the contact. The NRC will contact this individual if there are questions about the form. Notify the NRC of changes in the contact person on NRC Form 241.

7.5   Item 5: Office Telephone and Work Cell Numbers

Provide office and work cell telephone numbers for the individual who can answer questions about information submitted on NRC Form 241. Notify the NRC of changes in telephone numbers on NRC Form 241.

7.6   Item 6: Facsimile Number

Provide a facsimile number for the individual who can answer questions about NRC Form 241. Notify the NRC of changes in facsimile number on NRC Form 241.

7.7   Item 7: E-Mail Address

Provide the e-mail address of the individual who can answer questions about information submitted on NRC Form 241. Notify the NRC of changes in e-mail address on NRC Form 241. The e-mail address should not be a personal e-mail account.

7.8   Item 8: Activities To Be Conducted In Non-Agreement States Under the General License Given in 10 CFR 150.20

Check the appropriate box corresponding to the activities you propose to conduct under the general license. If the activity is not specifically identified, check the "other" box and provide a description of the activity as authorized by the Agreement State license. If you will be conducting industrial radiography, confirm that you have registered with the NRC as a package user as specified in 10 CFR 71.17(c)(3) and provide the packaging Certificate of Compliance numbers.

Limits on use

A legal entity that performs the same licensed activity under more than one Agreement State license may not exceed the 180-day limit for this activity in a calendar year. For example, a company authorized to conduct radiography under both a Texas license and a Mississippi license may conduct radiography in NRC jurisdiction, but its radiography activities in NRC jurisdiction will be limited to 180 days in a calendar year. However, a company holding two or more Agreement State licenses for different licensed activities may file one NRC Form 241 for each type of activity in any calendar year, allowing 180 days for each activity in NRC jurisdiction in a calendar year.

Certificate of Compliance Registration

Type B transportation packages, such as some radiography devices or overpacks used with radiography devices, are issued a Certificate of Compliance when the NRC approves the package. Before the licensee's first use of the package, the licensee must register with the NRC. A licensee engaging in radiography activities must register with the NRC as a user for each approved package issued a Certificate of Compliance, in accordance with the requirements of 10 CFR 71.17, "General License: NRC-Approved Package." The registration application must be sent to the Document Control Desk, Director, Division of Spent Fuel Storage and Transportation, Office of Nuclear Material Safety and Safeguards, U.S. Nuclear Regulatory Commission, Washington, DC 20555-0001. The information provided must include the licensee's name, license number, and the package identification number specified in the package approval. The supplier should provide the particular information to the licensee on request.

7.9   Item 9: Client Name, Address, City/County, State, ZIP Code

Provide the names of your clients and their mailing addresses.

7.10   Item 10: Actual Physical Address of Work Location

Any number of work locations may be listed on NRC Form 241. Specify the street address, city, and State or other descriptive address (e.g., nearest intersection, site name, building name or number, project name) for each work location. If the job site is linear and mobile (e.g., pipeline construction, multiple bridge sites), the descriptive address should include sufficient detail to allow an NRC inspector to find the work location, see Figure 7.1. A post office box address or general area locations are not acceptable. If known, provide Global Positioning System (GPS) coordinates and directions from a nearby recognizable location, such as a post office, police or fire station, shopping center, or municipal building. For offshore waters locations, see Figure 7.2, specify the oil field, block number, vessel name, platform, or laybarge.

Figure 7.2 Remote work location. Provide a descriptive address of remote work locations; e.g., well site in the Gulf of Mexico (specify block number, vessel name, platform, or name of the laybarge).

7.11 Item 11: Client Telephone Number

Provide office and cell telephone numbers for the client(s) identified in Item 9. The telephone numbers should be for an individual who is knowledgeable about the work to be conducted. Notify the NRC of changes in telephone numbers on NRC Form 241.

7.12 Item 12: Work Location Office Telephone and Cell Numbers

Provide office and cell telephone numbers at the work location for a licensee or client contact who can answer questions about the licensed activities at the site. Notify the NRC of changes in the work location telephone number on NRC Form 241.

7.13 Item 13: Dates Scheduled

Provide the date span identifying the start and end dates for each location. Indicate whether you plan to work on weekends.

For the initial filing, it is acceptable to indicate tentative dates, provided that you submit a change when adding or deleting dates as they become known. Changes should be submitted 3 days before beginning work activities within exclusive Federal jurisdiction. It is important that licensees track the days of use so that they do not exceed the 180-day limit.

> *Example:* The initial NRC Form 241 may list March 1 to March 31 for the site at the Bisco pipeline; however, because of rain, work was not performed on March 2 through March 10. The licensee should submit a change to delete these dates and receive credit for days not worked, because it may become important later in the calendar year should the work approach the 180-day limit.

7.14 Item 14: Number of Work Days

Provide the total number of days scheduled for use at the work location.

Reciprocity in each calendar year (January 1 through December 31) is limited to 180 days, except for activities in offshore waters. A "calendar day of use" is one on which you store or use licensed material in NRC jurisdiction. Licensed activities conducted on the same day at different locations within NRC jurisdiction count as 1 calendar day of use.

> *Example:* A licensee stores material at Site A on Monday and conducts licensed activities at Site A on Tuesday. The total is 2 days of use.

7.15 Item 15: Add

When submitting a change, if the dates specified in Item 13 ("Dates Scheduled") are to be added for a specified work location, indicate in this field the number of added days. The number of days may be the same as the entire interval or close to the actual number of work days.

7.16 Item 16: Delete

When submitting a change, if the dates specified in Item 13 ("Dates Scheduled") are to be deleted for a specified work location, indicate in this field the number of deleted days. You may provide a list of deleted work days.

7.17 Item 17: Location Reference Number

The NRC will generate a Location Reference Number for each work location in order to track reciprocity activities. An NRC-signed copy of NRC Form 241, which identifies the Location Reference Number corresponding to each work location, will be returned to the licensee.

Licensees should reference the appropriate Location Reference Number when submitting changes within the calendar year.

7.18 Item 18: List Radioactive Material, Which Will Be Possessed, Used, Installed, Serviced, or Tested

List radioactive material that will be possessed, stored, used, installed, serviced, or tested. Include a description of the type and quantity of radioactive material, sealed sources, or devices to be used. For sealed sources and devices, include the manufacturer and model number.

In accordance with 10 CFR Part 37, "Physical Protection of Category 1 and Category 2 Quantities of Radioactive Material," licensees that possess an aggregated Category 1 or Category 2 quantity of radioactive material must establish, implement, and maintain an access authorization program and a security program to ensure physical protection of the radioactive material. For additional guidance on the development of a 10 CFR Part 37 security program, see NUREG-2155, "Implementation Guidance for 10 CFR Part 37, "Physical Protection of Category 1 and Category 2 Quantities of Radioactive Material.")

When completing NRC Form 241, licensees should properly mark the Form if radioactive material exceeds the threshold levels in 10 CFR Part 37.

Sensitive security-related information in a Reciprocity request includes types and quantities of byproduct material and should be marked: "Security Related—Withhold under 10 CFR 2.390." For further information, see RIS 2005-31, "Control of Security-Related Sensitive Unclassified Non-Safeguards Information Handled by Individuals, Firms, and Entities Subject to NRC Regulation of the Use of Source, Byproduct, and Special Nuclear Material," dated December 22, 2005, which can be found on the NRC's Generic Communications webpage under Regulatory Issue Summaries: http://www.nrc.gov/reading-rm/doc-collections/gen-comm/.

Example: Portable moisture/density gauge; manufacturer; model number; americium-241, 296 megabecquerels (9 millicuries (mCi)); cesium-137, 370 megabecquerels (4 mCi).

Note: In accordance with the guidance in RIS 2005-31, this example would not require the "Security Related Information—Withhold under 10 CFR 2.390," marking. See Chapter 8, "Identifying and Protecting Sensitive Information," for more information about marking documents.

Please contact the appropriate regional office for questions regarding the security of licensed material or applicability of 10 CFR Part 37.

7.19 Item 19: Agreement State Specific License Which Authorizes the Undersigned To Conduct Activities Which Are the Same Except for Location of Use

License Number: Provide the Agreement State license number that authorizes the activity.

State: Name the Agreement State that issued the license.

Expiration Date: Provide the expiration date of the Agreement State license.

> **Note:** This note reminds licensees that if the license expiration date is in the current calendar year, the licensee may work under the license until the expiration date. No work in NRC jurisdiction may be done after the expiration date, unless the Agreement State has provided a copy of the Deemed Timely Letter, indicating that a renewal is pending and that the Agreement State license continues in effect. When the Agreement State authority has completed action on the renewal, a copy of the license shall be forwarded to the NRC regional office. The NRC may contact the Agreement State licensing authority to confirm that the Agreement State license is current and in effect.

7.20 Item 20: Certification

Print or type the name and title of your certifying officer (e.g., radiation safety officer or management representative). This individual should sign and date NRC Form 241.

8. IDENTIFYING AND PROTECTING SENSITIVE INFORMATION

All licensing applications, except for portions containing sensitive information, will be made available for review in the NRC's Public Document Room and electronically at the NRC Library. For more information on the NRC Library, visit www.nrc.gov.

The licensee should identify, mark, and protect sensitive information against unauthorized disclosure to the public. Licensing applications that contain sensitive information should be marked as indicated below in accordance with 10 CFR 2.390 before the information is submitted to the NRC. Key examples are as follows:

- Proprietary Information/Trade Secrets: If it is necessary to submit proprietary information or trade secrets, follow the procedure in 10 CFR 2.390(b). Failure to follow this procedure could result in disclosure of the proprietary information to the public or substantial delays in processing the application.

- Personally Identifiable Information: Personally identifiable information (PII) about employees or other individuals should not be submitted unless specifically requested by the NRC. Examples of PII are social security number, home address, home telephone number, date of birth, and radiation dose information. If PII is submitted, a cover letter should clearly state that the attached documents contain PII and the top of every page of a document that contains PII should be clearly marked as follows: "Privacy Act Information—Withhold Under 10 CFR 2.390." For further information, see Regulatory Issue Summary (RIS) 2007-04, "Personally Identifiable Information Submitted to the U.S. Nuclear Regulatory Commission," dated March 9, 2007, which can be found on the NRC's Generic Communications webpage under "Regulatory Issue Summaries": http://www.nrc.gov/reading-rm/doc-collections/gen-comm/.

- Security-Related Information: Following the events of September 11, 2001, the NRC changed its procedures to avoid release of information that terrorists could use to plan or execute an attack against facilities or citizens in the United States. As a result, certain types of information are no longer routinely released and are treated as sensitive unclassified information. For example, certain information about the quantities and locations of radioactive material at licensed facilities, and associated security measures, are no longer released to the public. Therefore, a cover letter should clearly state that the attached documents contain sensitive security-related information and the top of every page of a document that contains such information should be clearly marked: "Security Related Information—Withhold under 10 CFR 2.390." For the pages having security-related sensitive information, an additional marking should be included (e.g. an editorial note box) adjacent to that material. For further information, see RIS 2005-31, "Control of Security-Related Sensitive Unclassified Non-Safeguards Information Handled by Individuals, Firms, and Entities Subject to NRC Regulation of the Use of Source, Byproduct, and Special Nuclear Material," dated December 22, 2005, which can be found on the NRC's Generic Communications webpage under "Regulatory Issue Summaries": http://www.nrc.gov/reading-rm/doc-collections/gen-comm/. Additional information on procedures and any updates is available at http://www.nrc.gov/reading-rm/sensitive-info.html.

The NRC recognizes that certain information reported on NRC Form 241 may be sensitive and that the public release of such information may have an adverse impact on the licensee. Licensees wishing the NRC to withhold from public disclosure, as proprietary the information contained in Items 9 to 13 of NRC Form 241 should submit an application for withholding, accompanied by an affidavit. The application and affidavit must be submitted in accordance with 10 CFR 2.390(b). Failure to follow this procedure may result in disclosure of the proprietary information to the public or substantial delays in processing NRC Form 241.

The regulations list various forms of information that can be protected from public disclosure. These include:

- Trade secrets and commercial or financial information;

- Interagency or intragency memoranda or letters that would not be available by law to a party other than an agency in litigation with NRC;

- Certain records or information compiled for law enforcement purposes;

- Geological and geophysical information and data, including maps, or information concerning wells;

- Personnel, medical, or other information, the disclosure of which would constitute a clearly unwarranted invasion of personal privacy.

In 10 CFR 2.390, the NRC specifies the procedures and requirements for persons to submit sensitive information to the NRC so that it may be properly protected from disclosure. This regulation is available electronically on the Commission's website: www.nrc.gov/reading-rm/doc-collections/cfr.

Except for personal privacy information, which is not subject to the affidavit requirement, if the NRC determines that the application or affidavit is deficient (i.e., does not contain the required information as outlined in 10 CFR 2.390), the applicant will be notified that additional information is needed and that the review will continue when the required information is received.

If the request is denied, in whole or in part, the NRC will give the applicant the option of withdrawing the information or application, as permitted in 10 CFR 2.390. If the applicant decides not to withdraw the information or application, the NRC will notify the applicant in writing that the request for withholding has been denied and that the NRC will disregard any references concerning the proprietary status of the information.

Any part of NRC Form 241 that the NRC has determined should be withheld from public disclosure will be handled in accordance with Management Directive 12.6, "NRC Sensitive Unclassified Information Security Program," dated December 20, 1999, and the applicant will be notified in writing that the NRC plans to honor the request.

8.1 Response from the Licensee or Applicant

For the NRC to determine whether the information should be withheld from public disclosure, the licensee or applicant should provide the following information in sufficient explanatory detail, in addition to the other requirements of 10 CFR 2.390:

- A statement requesting that a document be withheld in whole or in part from public disclosure on the grounds that it contains sensitive information.
- A notarized affidavit that:
  — Identifies the document or part sought to be withheld and the position of the person making the affidavit;
  — Contains a full statement of the reasons why it is claimed that the information should be withheld from public disclosure. Such statement must include information about whether any of the following apply:

  The information has been held in confidence by its owner;

  - The information is of a type customarily held in confidence by its owner;

  - The information was transmitted to and received by the NRC in confidence;

  - The information is available in public sources;

  - Public disclosure of the information sought to be withheld is likely to cause substantial harm to the competitive position of the owner of the information, taking into account the value of the information to the owner; the amount of effort or money, if any, expended by the owner in developing the information; and the ease or difficulty with which the information could be properly acquired or duplicated by others.

An affidavit submitted by a company should be executed by an officer or upper-level management official. This individual must have been specifically delegated the function of reviewing the information sought to be withheld and be authorized to apply for its withholding on behalf of the company. The affidavit must be executed by the owner of the information, even if the information sought to be withheld is submitted by another person.

The statement and affidavit should be submitted at the same time that the information sought to be withheld is filed with the NRC. The information sought to be withheld should be detailed separately, with any information specified in the affidavit as a trade secret, confidential, or privileged commercial or financial information so indicated. If an NRC Form 241 has already been submitted for the calendar year, the licensee must submit an application for withholding and an affidavit within 3 days of the NRC's receipt of Form 241. Once the NRC agrees with the request for withholding, the request will be maintained as valid for as long as the Agreement State licensee continues to perform reciprocity activities.

If the licensee skips a year between filing reciprocity reports, the application and affidavit for withholding must be resubmitted for review.

Appendix D includes a checklist for requests for withholding information from public disclosure.

Anyone submitting a request to withhold information from public disclosure should thoroughly review 10 CFR 2.390 and be familiar with its requirements and limitations.

Withholding from public inspection shall not affect the right, if any, of persons properly and directly concerned to inspect the documents. If the need arises, the NRC may send copies of this information to NRC consultants working in that area. The NRC will ensure that the consultants have signed the appropriate agreements for handling proprietary information.

If the basis for withholding this information from public inspection should change in the future such that the information could then be made available for public inspection, the applicant should promptly notify the NRC. The applicant also should understand that the NRC may have cause to review this determination in the future; for example, if the scope of a Freedom of Information Act request includes the information in question. In all review situations, if the NRC makes a determination adverse to the above, the applicant will be notified in advance of any public disclosure. Anyone submitting commercial or financial information they believe to be privileged, confidential, or a trade secret must remember that the NRC's policy is to achieve an effective balance between legitimate concerns for the protection of competitive positions and the right of the public to be fully apprised of the basis for, and the effects of, licensing or rulemaking actions. It is within the NRC's discretion to withhold such information from public disclosure.

The NRC staff should review and follow the procedures in Appendix E when handling information that a licensee or applicant has requested to be withheld from public disclosure.

APPENDIX A

U.S. NUCLEAR REGULATORY COMMISSION FORM 241

Please use the most current version of this form, which may be found at:
http://www.nrc.gov/reading-rm/doc-collections/forms/nrc241.pdf

NRC FORM 241 (03-2013)	U.S. NUCLEAR REGULATORY COMMISSION	APPROVED BY OMB: NO. 3150-0013 EXPIRES: (10/31/2014)

Estimated burden per response to comply with this mandatory collection request: 30 minutes. This notification is required so that NRC may schedule inspection of the activities to ensure that they are conducted in accordance with requirements for protection of the public health and safety. Send comments regarding burden estimate to the Information Services Branch (T-5 F53), U. S. Nuclear Regulatory Commission, Washington, DC 20555-0001 or by internet e-mail to infocollects.Resource@nrc.gov, and to the Desk Officer, Office of Information and Regulatory Affairs, NEOB-10202, (3150-0013), Office of Management and Budget, Washington, DC 20503. If a means used to impose an information collection does not display a currently valid OMB control number, the NRC may not conduct or sponsor, and a person is not required to respond to, the information collection.

REPORT OF PROPOSED ACTIVITIES IN NON-AGREEMENT STATES, AREAS OF EXCLUSIVE FEDERAL JURISDICTION, OR OFFSHORE WATERS
(Please read the instructions before completing this form)

1. NAME OF LICENSEE *(Person or firm proposing to conduct the activities described below)*

2. TYPE OF REPORT
☐ INITIAL ☐ CHANGE

3. ADDRESS OF LICENSEE *(Mailing address or other location where licensee may be located)*

4. LICENSEE CONTACT AND TITLE

5. Office Number | 5a. Work Cell Number | 6. FACSIMILE NUMBER

7. EMAIL ADDRESS

8. ACTIVITIES TO BE CONDUCTED UNDER THE GENERAL LICENSE GIVEN IN 10 CFR 150.20

☐ WELL LOGGING ☐ LEAK TESTING AND/OR CALIBRATIONS ☐ TYPE OF SERVICE _____

☐ PORTABLE GAUGES ☐ OTHER (Specify) ⇒ _____

☐ RADIOGRAPHY ⇒ REGISTERED AS USER OF PACKAGING (CERTIFICATES OF COMPLIANCE NUMBERS) _____

9. CLIENT NAME, ADDRESS, CITY/COUNTY, STATE, ZIP CODE

10. ACTUAL PHYSICAL ADDRESS OF WORK LOCATION *(Street and Number or other location. Provide GPS coordinates if known.)*

11. CLIENT TELEPHONE NUMBER
Office | Work Cell

12. WORK LOCATION TELEPHONE NUMBER
Office | Work Cell

13. DATES SCHEDULED			14. NUMBER OF WORK DAYS	15. ADD	16. DELETE	17. LOCATION REFERENCE NUMBER
FROM	TO	WEEKENDS ☐ YES ☐ NO				NUMBER TO BE ASSIGNED BY NRC

LIST ADDITIONAL WORK SITES ON SEPARATE SHEET(S) TO INCLUDE ALL INFORMATION CONTAINED IN ITEMS 10-17 ABOVE.

18. LIST RADIOACTIVE MATERIAL, WHICH WILL BE POSSESSED, USED, INSTALLED, SERVICED, OR TESTED *(Include description of type and quantity of radioactive material, sealed sources, or devices to be used.)*

Device Type/Sealed Sources

Model No.

19. AGREEMENT STATE SPECIFIC LICENSE WHICH AUTHORIZES THE UNDERSIGNED TO CONDUCT ACTIVITIES WHICH ARE THE SAME, EXCEPT FOR LOCATION OF USE, AS SPECIFIED IN ITEM 10, ABOVE. *(One copy of the specific license must accompany the initial NRC Form 241.)*

LICENSE NUMBER | STATE | EXPIRATION DATE

20. CERTIFICATION (MUST BE COMPLETED BY APPLICANT)

I, THE UNDERSIGNED, HEREBY CERTIFY THAT:

a. All information in this report is true and complete.

b. I have read and understand the provision of the general license 10 CFR 150.20 reprinted on the instructions of this form; and I understand that I am required to comply with these provisions as to all byproduct, source, or special nuclear material which I possess and use in non-Agreement States or offshore waters under the general license for which this report is filed with the U. S. Nuclear Regulatory Commission.

c. I understand that activities, including storage, conducted in non-Agreement States under general license 10 CFR 150.20 are limited to a total of 180 days in calendar year. With the exception of work conducted in off-shore waters, which is authorized for an unlimited period of time in the calendar year.

d. I understand that I may be inspected by NRC at the above listed work site locations and at the Licensee home office address for activities performed in non-Agreement States or offshore waters.

e. I understand that conduct of any activities not described above, including conduct of activities on dates or locations different from those described above or without NRC authorization, may subject me to enforcement action, including civil or criminal penalties.

APPROVED BY (Printed Name and Title) | SIGNATURE | DATE | TOTAL USAGE - DAYS TO DATE

WARNING: False statements in this certificate may be subject to civil and/or criminal penalties. NRC regulations require that submissions to the NRC be complete and accurate in all material respects. 18 U.S.C. Section 1001 makes it a criminal offense to make a willfully false statement or representation to any department or agency of the United States as to any matter within its jurisdiction.

FOR NRC USE ONLY

APPROVED BY *(Typed/Printed Name and Title)* | SIGNATURE | DATE | TOTAL USAGE -- DAYS TO DATE

☐ NON-PUBLIC ☐ SENSITIVE_SECURITY RELATED, MD 3.4 Non-Public a.3 ADAMS ML#

NRC FORM 241 (03-2013)

A-1

APPENDIX B

**RECOMMENDED PROCEDURE TO OBTAIN JURISDICTION
DETERMINATIONS FOR FEDERAL SITES**

If you are uncertain about the jurisdictional status of the proposed job site, the NRC recommends that you take the following steps:

- Obtain specific information about the location of the proposed work site (e.g., street address, range or township, building or hangar number, distance from a specific intersection, or other identifying details) and the identity of the Federal agency controlling the proposed work site.

- Call the Federal agency's local contact (e.g., contract officer, base environmental health officer, judge advocate general, district office staff, regional office staff, regional counsel) and request information about the jurisdictional status of the proposed work site. It is recommended that licensees request such a statement of jurisdiction in writing. Otherwise, licensees should document for their records the name and title of the person at the Federal agency who provided the determination and the date that it was provided.

- In an Agreement State if the work site is identified as falling under "exclusive Federal jurisdiction" or within the protected or controlled access area of an NRC licensed facility and the licensee is an Agreement State licensee, submit the notification of proposed work (U.S. Nuclear Regulatory Commission (NRC) Form 241, "Report of Proposed Activities in Non-Agreement States, Areas of Exclusive Federal Jurisdiction, or Offshore Waters") and, if available, a copy of the statement of jurisdiction from the Federal agency to the appropriate NRC regional office. In lieu of submitting NRC Form 241, Agreement State licensees may apply for a specific NRC license to operate in areas under NRC jurisdiction. If the licensee is an NRC licensee with authorization to work at temporary job sites, no additional action is required.

- If the work site is identified as other than "exclusive Federal jurisdiction," the licensee should contact the Agreement State's Radiation Control Program Office within which the facility is located for information about that State's regulations.

- In an Agreement State, if you are a company proposing to work on Tribal lands, you should contact the appropriate NRC regional office to help determine whether the NRC or the Agreement State has regulatory authority.

- Questions about Federal jurisdiction determinations should be referred to the NRC Regional office identified in Figure 2.2

Appendix C

**U.S. NUCLEAR REGULATORY COMMISSION FORM 629—
AUTHORIZATION FOR PAYMENT BY CREDIT CARD**

Please use the most current version of this form, which may be found at:
http://www.nrc.gov/reading-rm/doc-collections/forms/nrc629.pdf

NRC FORM 629 (6-2011)	U.S. NUCLEAR REGULATORY COMMISSION	APPROVED BY OMB: NO. 3150-0190	EXPIRES: 06/30/2014
AUTHORIZATION FOR PAYMENT BY CREDIT CARD		Estimated burden per response to comply with this voluntary collection request: 5 minutes. Requested information will allow respondents to transfer funds electronically. Send comments regarding burden estimate to the Information Services Branch (T-5 F53), U.S. Nuclear Regulatory Commission, Washington, DC 20555-0001, or by internet e-mail to infocollects.resource@nrc.gov, and to the Desk Officer, Office of Information and Regulatory Affairs, NEOB-10202, (3150-0190), Office of Management and Budget, Washington, DC 20503. If a means used to impose an information collection does not display a currently valid OMB control number, the NRC may not conduct or sponsor, and a person is not required to respond to, the information collection.	

The NRC is currently accepting credit card payment of fees and other debts. If you wish to pay by credit card, complete the authorization below. If you have any questions, contact NRC's collection service provider, Department of Interior/National Business Center at (303) 969-5880.

NAME OF CARDHOLDER	ACCOUNT NUMBER	CARD EXPIRATION DATE
ADDRESS	CARDS ACCEPTED (Check card you are using) ☐ VISA ☐ MASTERCARD ☐ NOVUS (DISCOVER) ☐ AMERICAN EXPRESS	
TELEPHONE LICENSE NUMBER	SIGNATURE OF CARDHOLDER	
INVOICE NUMBER OR DESCRIPTION TOTAL AMOUNT OF TRANSACTION		

If you are paying an application or registration fee (including new licenses, amendments, etc.), mail the completed form with your application to the appropriate regional office.

For all other payments, send completed form to:

U.S. Bank
U.S. Nuclear Regulatory Commission
Accounts Receivable Team
P. O. Box 979051
St. Louis, MO 63197-9000

PRIVACY ACT STATEMENT

Pursuant to 5 U.S.C. 552a(e)(3), enacted into law by Section 3 of the Privacy Act of 1974 (Public Law 93-579), the following statement is furnished to individuals who supply information to the U.S. Nuclear Regulatory Commission (NRC) on NRC Form 629. This information is maintained as part of a system of records designated as NRC-32 and described at 75 Federal Register 57356 (September 20, 2010), or the most recent Federal Register publication of the NRC's "Republication of Systems of Records Notices" that is available in the NRC's Agencywide Documents Access and Management System.

1. **AUTHORITY:** 5 U.S.C. 552a; 5 U.S.C. 5514; 15 U.S.C. 1681; 26 U.S.C. 6103; 31 U.S.C. Chapter 37; 31 U.S.C. 6501-6508; 42 U.S.C. 2201; 42 U.S.C. 5841; 31 CFR 900-904; 10 CFR Parts 15, 16, 170, 171; Executive Order (E.O.) 9397, as amended by E.O. 13478; and E.O. 12731.

2. **PRINCIPAL PURPOSE(S):** To collect credit card account information and provide authorization for its use to collect a payment or debt.

3. **ROUTINE USE(S):** Information contained in this system may be disclosed to debt collection contractors or to other Federal agencies for the purpose of collecting and reporting on delinquent debts and to banks enrolled in the Automated Clearinghouse (ACH) Network to collect a payment or debt when the individual has given his or her authorization for this purpose. Information may be disclosed in accordance with any of the Routine Uses listed in the Prefatory Statement of General Routine Uses, including to an appropriate Federal, State, local or Foreign agency in the event the information indicates a violation or potential violation of law; in the course of an administrative or judicial proceeding; to an appropriate Federal, State, local and Foreign agency to the extent relevant and necessary for an NRC decision about you or to the extent relevant and necessary for that agency's decision about you; in the course of discovery under a protective order issued by a court of competent jurisdiction and in presenting evidence; to a Congressional office to respond to their inquiry made at your request; to NRC-paid experts, consultants, and others under contract with the NRC, on a need-to-know basis; and to appropriate persons and entities for purposes of response and remedial efforts in the event of a suspected or confirmed breach of data from this system of records.

4. **WHETHER DISCLOSURE IS MANDATORY OR VOLUNTARY AND EFFECT ON INDIVIDUAL OF NOT PROVIDING INFORMATION:** Providing this information is voluntary. However, not providing the requested information will not provide the NRC the information and authorization required to use your credit card to collect a payment or debt.

5. **SYSTEM MANAGER(S) AND ADDRESS:** Controller, Division of the Controller, Office of the Chief Financial Officer, U.S. Nuclear Regulatory Commission, Washington, DC 20555-0001.

NRC FORM 629 (6-2011)

APPENDIX D

CHECKLIST FOR REQUESTS TO WITHHOLD INFORMATION FROM PUBLIC DISCLOSURE (UNDER 10 CFR 2.390)

In order to request that the U.S. Nuclear Regulatory Commission (NRC) withhold information contained in an NRC Form 241, "Report of Activities in Non-Agreement States, Areas of Exclusive Federal Jurisdiction, or Offshore Waters," from public disclosure, the applicant must submit the information and NRC Form 241, including an affidavit, in accordance with Title 10 of the *Code of Federal Regulations* (10 CFR) 2.390, "Public Inspections, Exemptions, Requests for Withholding." The applicant should submit all of the following:

☐	A proprietary copy of the information. Brackets should be placed around the material considered to be proprietary. This copy should be marked as proprietary.
☐	A non-proprietary copy of the information. Applicants should white out or black out the proprietary portions (i.e., those in the brackets), leaving the non-proprietary portions intact. This copy should **not** be marked as proprietary.
☐	An affidavit that:
☐	Is notarized.
☐	Clearly identifies (such as by name or title and date) the document to be withheld.
☐	Clearly identifies the position of the person executing the affidavit. This person must be an officer or upper-level management official who has been delegated the function of reviewing the information the company is seeking to withhold and is authorized to apply for withholding on behalf of the company.
☐	States that the company submitting the information is the owner of the information or is required, by agreement with the owner of the information, to treat the information as proprietary.
☐	Provides a rational basis for holding the information in confidence.
☐	Fully addresses the following issues:
☐	Is the information submitted to, and received by, the NRC in confidence? Provide details.
☐	To the best of the applicant's knowledge, is the information currently available in public sources?
☐	Does the applicant customarily treat this information, or this type of information, as confidential? Explain why.
☐	Would public disclosure of the information be likely to cause substantial harm to the competitive position of the applicant? If so, explain why in detail. The explanation should include the value of the information to your company, the amount of effort or money expended in developing the information, and the ease or difficulty for others to acquire the information.

APPENDIX E

PROCEDURES FOR NRC PERSONNEL REGARDING REQUESTS TO WITHHOLD INFORMATION FROM PUBLIC DISCLOSURE (UNDER 10 CFR 2.390)

In Title 10 of the *Code of Federal Regulations* (10 CFR) 2.390, "Public inspections, exemptions, requests for withholding," the U.S. Nuclear Regulatory Commission (NRC) permits individuals submitting documents to the NRC to request that trade secrets or privileged or confidential commercial or financial information in those documents be withheld from public disclosure. (Refer to 10 CFR 2.390(a)(4)). Trade secrets and commercial or financial information deemed privileged or confidential by the submitter are commonly known in the aggregate as "proprietary information," although the regulation does not use that term.

A request for withholding proprietary information from public disclosure should be reviewed in accordance with established regional instructions, which provide guidance to regional personnel on the review of a request to withhold proprietary information from public disclosure.

The regulation at 10 CFR 2.390 requires a person who proposes that a document be withheld from public disclosure on the grounds that it contains proprietary information to submit an application for withholding accompanied by an affidavit that provides the reasons for the proposed withholding (refer to 10 CFR 2.390(b)(1)(iii)).

Information relating to requests for withholding information from public disclosure on other grounds, such as personal privacy reasons, can be found at 10 CFR 2.390(a).

The Commission has directed that internal procedures be created to ensure that the NRC staff expeditiously determines whether a request for non-disclosure of proprietary information shall be granted. Only those portions of a document containing proprietary information may be withheld from public disclosure under these procedures; non-proprietary portions may not be withheld from public disclosure unless a separate basis for withholding applies. The staff also should ascertain promptly whether the submitter would like a document returned, if possible, in those cases where the agency denies the request to withhold proprietary information from public disclosure.

Upon receipt of a document requested to be withheld from public disclosure as containing proprietary information or proprietary in whole, the NRC staff will promptly determine, in consultation with the regional counsel, whether the justification provided by the submitter in its affidavit supports a finding that the information sought to be withheld is proprietary and thus should be withheld from public disclosure. If an affidavit has not been provided with the document, an affidavit from the owner shall be submitted pursuant to 10 CFR 2.390(b)(1) before a determination is made on whether to withhold the information. Again, it is important, when reviewing the request and supporting affidavit, to bear in mind that the presence of proprietary information in a document does not justify withholding the entire document if non-proprietary information can be reasonably segregated from proprietary information. If the staff determines that some information or the entire document submitted to the NRC proprietary, the staff should prepare a written response to the person who has requested non-disclosure stating that the proprietary information or document, as appropriate, will be withheld from public disclosure on the grounds that the withheld information or document constitutes trade secrets or commercial or financial information deemed privileged or confidential. (Refer to 10 CFR 2.390(a)(4)).

APPENDIX F

SAFETY CULTURE STATEMENT OF POLICY

Safety Culture

The safety culture policy statement was published in the *Federal Register* (76 FR 34773) on June 14, 2011, and can be found at: http://www.gpo.gov/fdsys/pkg/FR-2011-06-14/pdf/2011-14656.pdf. It is also posted in the NRC's Agencywide Documents Access and Management System (ADAMS) and can be found using Accession Number ML11146A047.

Safety Culture Policy Statement

The purpose of this Statement of Policy is to set forth the Commission's expectation that individuals and organizations establish and maintain a positive safety culture commensurate with the safety and security significance of their activities and the nature and complexity of their organizations and functions. This includes all licensees, certificate holders, permit holders, authorization holders, holders of quality assurance program approvals, vendors and suppliers of safety-related components, and applicants for a license, certificate, permit, authorization, or quality assurance program approval, subject to NRC authority. The Commission encourages the Agreement States, Agreement State licensees and other organizations interested in nuclear safety to support the development and maintenance of a positive safety culture, as articulated in this Statement of Policy.

Nuclear Safety Culture is defined as *the core values and behaviors resulting from a collective commitment by leaders and individuals to emphasize safety over competing goals to ensure protection of people and the environment.* Individuals and organizations performing regulated activities bear the primary responsibility for safety and security. The performance of individuals and organizations can be monitored and trended and, therefore, may be used to determine compliance with requirements and commitments and may serve as an indicator of possible problem areas in an organization's safety culture. The NRC will not monitor or trend values. These will be the organization's responsibility as part of its safety culture program.

Organizations should ensure that personnel in the safety and security sectors have an appreciation for the importance of each, emphasizing the need for integration and balance to achieve both safety and security in their activities. Safety and security activities are closely intertwined. While many safety and security activities complement each other, there may be instances in which safety and security interests create competing goals. It is important that consideration of these activities be integrated so as not to diminish or adversely affect either; thus, mechanisms should be established to identify and resolve these differences. A safety culture that accomplishes this would include all nuclear safety and security issues associated with NRC-regulated activities.

Experience has shown that certain personal and organizational traits are present in a positive safety culture. A trait, in this case, is a pattern of thinking, feeling, and behaving that emphasizes safety, particularly in goal conflict situations, e.g., production, schedule, and the cost of the effort versus safety. It should be noted that although the term "security" is not expressly included in the following traits, safety and security are the primary pillars of the NRC's regulatory mission. Consequently, consideration of both safety and security issues, commensurate with their significance, is an underlying principle of this Statement of Policy.

The following are traits of a positive safety culture:

(1)   *Leadership Safety Values and Actions* – Leaders demonstrate a commitment to safety in their decisions and behaviors;

(2)   *Problem Identification and Resolution* – Issues potentially impacting safety are promptly identified, fully evaluated, and promptly addressed and corrected commensurate with their significance;

(3)   *Personal Accountability* – All individuals take personal responsibility for safety;

(4)   *Work Processes* – The process of planning and controlling work activities is implemented so that safety is maintained;

(5)   *Continuous Learning* – Opportunities to learn about ways to ensure safety are sought out and implemented;

(6)   *Environment for Raising Concerns* – A safety-conscious work environment is maintained where personnel feel free to raise safety concerns without fear of retaliation, intimidation, harassment, or discrimination;

(7)   *Effective Safety Communication* – Communications maintain a focus on safety;

(8)   *Respectful Work Environment* – Trust and respect permeate the organization; and

(9)   *Questioning Attitude* – Individuals avoid complacency and continuously challenge existing conditions and activities in order to identify discrepancies that might result in error or inappropriate action.

There may be traits not included in this Statement of Policy that are also important in a positive safety culture. It should be noted that these traits were not developed to be used for inspection purposes.

It is the Commission's expectation that all individuals and organizations, performing or overseeing regulated activities involving nuclear materials, should take the necessary steps to promote a positive safety culture by fostering these traits as they apply to their organizational environments. The Commission recognizes the diversity of these organizations and acknowledges that some organizations have already spent significant time and resources in the development of a positive safety culture. The Commission will take this into consideration as the regulated community addresses the Statement of Policy.

NRC FORM 335 (12-2010) NRCMD 3.7	U.S. NUCLEAR REGULATORY COMMISSION **BIBLIOGRAPHIC DATA SHEET** *(See instructions on the reverse)*	1. REPORT NUMBER (Assigned by NRC, Add Vol., Supp., Rev., and Addendum Numbers, if any.) NUREG-1556, Volume 19, Revision 1 DRAFT

2. TITLE AND SUBTITLE	3. DATE REPORT PUBLISHED	
Consolidated Guidance about Materials Licenses: Guidance for Agreement State Licensees about NRC Form 241 "Report of Proposed Activities in Non-Agreement States, Areas of Exclusive Federal Jurisdiction, or Offshore Waters" and Guidance for NRC Licensees Proposing to Work in Agreement State Jurisdiction (Reciprocity) (NUREG-1556, Volume 19, Revision 1) - Draft Report for Comment	MONTH August	YEAR 2013
	4. FIN OR GRANT NUMBER	

5. AUTHOR(S)	6. TYPE OF REPORT
Leira Cuadrado, Craig Z. Gordon, Michelle R. Simmons, Geoffrey M. Warren	Technical
	7. PERIOD COVERED (Inclusive Dates) January 1999 to June 2013

8. PERFORMING ORGANIZATION - NAME AND ADDRESS (If NRC, provide Division, Office or Region, U. S. Nuclear Regulatory Commission, and mailing address; if contractor, provide name and mailing address.)

Division of Materials Safety and State Agreements
Office of Federal and State Materials and Environmental Management Programs
U.S. Nuclear Regulatory Commission
Washington, DC 20555-0001

9. SPONSORING ORGANIZATION - NAME AND ADDRESS (If NRC, type "Same as above", if contractor, provide NRC Division, Office or Region, U. S. Nuclear Regulatory Commission, and mailing address.)

Same as above

10. SUPPLEMENTARY NOTES

11. ABSTRACT (200 words or less)

This technical report contains information intended to provide program-specific guidance and assist applicants and licensees in preparing applications for a general license under Title 10 of the Code of Federal Regulations (10 CFR) 150.20, "Recognition of Agreement State Licenses," by describing the types of information needed from the licensee to complete NRC Form 241, "Report of Proposed Activities in Non-Agreement States, Areas of Exclusive Federal Jurisdiction, or Offshore Waters." This report should be used in preparing requests for NRC Form 241; however, the guidance contained in this report does not represent new or proposed regulatory requirements. It is intended for use by Agreement State licensees, NRC licensees, and NRC staff, and it will also be available to Agreement States. This document also provides contact organization guidance to NRC licensees who wish to work in Agreement States.

12. KEY WORDS/DESCRIPTORS (List words or phrases that will assist researchers in locating the report.)	13. AVAILABILITY STATEMENT
NUREG-1556 Volume 19 Reciprocity 10 CFR Part 150.20	unlimited
	14. SECURITY CLASSIFICATION
	(This Page) unclassified
	(This Report) unclassified
	15. NUMBER OF PAGES
	16. PRICE

UNITED STATES
NUCLEAR REGULATORY COMMISSION
WASHINGTON, DC 20555-0001

OFFICIAL BUSINESS

NUREG-1556, Vol. 19
Revision 1, Draft

Consolidated Guidance about Materials Licenses: Guidance for Agreement State Licensees about NRC Form 241 "Report of Proposed Activities in Non-Agreement States, Areas of Exclusive Federal Jurisdiction, or Offshore Waters" and Guidance for NRC Licensees Proposing to Work in Agreement State Jurisdiction (Reciprocity)

August 2013